Fleas

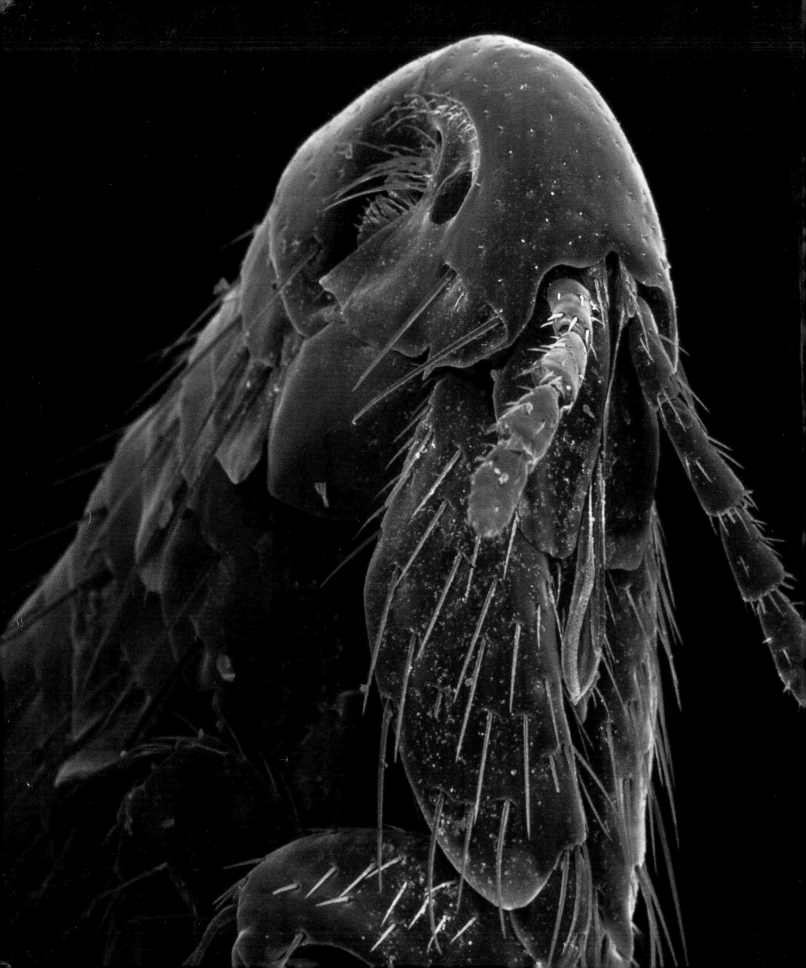

Fleas

Kathryn Stevens

THE CHILD'S WORLD®, INC.

Library of Congress Cataloging-in-Publication Data
Stevens, Kathryn, 1954–
Fleas / by Kathryn Stevens.
        p.   cm.
Includes index.
Summary: Describes the physical characteristics,
behavior, habitat, and life cycle of fleas.
ISBN 1-56766-632-9 (lib. bdg. : alk. paper)
1. Fleas—Juvenile literature.
[1. Fleas.]   I. Title.
        QL599.5.S74   1999
595.77'5—dc21   98-47584
                CIP
                AC

Photo Credits

ANIMALS ANIMALS © Bernard, G.I. OSF: 15
ANIMALS ANIMALS © George Bernard: 19, 20
ANIMALS ANIMALS © Stephen Dalton: 16
© Archive Photos: 26
© Dr. H.C. Robinson/Science Photo Library, The National Audubon Society Collection/Photo Researchers: 24
© Dwight R. Kuhn: cover, 9, 30
© John Kaprielian, The National Audubon Society Collection/Photo Researchers: 6
© Kent & Donna Dannen: 29
© K.H. Kjeldsen/Science Photo Library, The National Audubon Society Collection/Photo Researchers: 23
© Kim Taylor/Bruce Coleman, Inc.: 10, 13
© Oliver Meckes, The National Audubon Society Collection/Photo Researchers: 2

On the cover...

Front cover: This flea is jumping above the fur of its host.
Page 2: This picture shows what a *human flea* looks like at 175 times its real size.

Table of Contents

It is a hot summer day. An old dog is snoozing on the front porch. The dog's paws twitch as she dreams. Perhaps she is dreaming of chasing rabbits! Suddenly the dog sits up, wide awake. Her back foot starts scratching and scratching at a spot near her collar. She has a brand-new, itchy bug bite. What pesky bug has disturbed the dog's nap? It's a flea!

⇐ This dog is scratching an itchy flea bite.

What Are Fleas?

Fleas are insects that live on warm-blooded animals. They hop around on the animals, biting their skin and sucking their blood. Creatures that feed on other living creatures are called **parasites. Hosts** are the animals on which the parasites live. Many birds and mammals all over the world, including people, are hosts for fleas.

This flea is moving through its host's fur. ⇒

What Do Fleas Look Like?

Fleas are strange-looking creatures. Unlike most insects, they have no wings. Instead, they have large, strong hind legs that are specially made for jumping. Fleas are very tiny. Most of them are less than an eighth of an inch long. Some are only four-hundredths of an inch long!

⇐ This *gray squirrel flea* is holding on to a single hair.

Because their bodies are flattened from side to side, fleas move easily through feathers or hair. A flea's body is hard and surprisingly strong. In fact, if you catch a flea, you cannot kill it by squishing it between your fingers! Fleas are well adapted to living and feeding on other animals. Their mouths are specially made for biting an animal's skin and sucking its blood.

This *bird flea* is crawling on a feather. ⇒

Are There Different Kinds of Fleas?

There are about 16,000 different kinds, or **species,** of fleas in the world. About 250 species live in North America. Many fleas are named for their favorite host animal. *Human fleas* like to live on people. *Rat fleas, cat fleas,* and *dog fleas* prefer to live on rats, cats, and dogs.

Some fleas have unusual habits. The *chigoe* lives in very warm regions. Female chigoes burrow into people's and animals' feet and lay eggs, causing painful bumps and sores. *Sticktight fleas* bite chickens and turkeys and hang on. Sometimes hundreds of sticktights fasten themselves to a single bird.

This *rabbit flea* is crawling in the fur of a rabbit's ear. ⇒

Fleas can leap a foot or more—an amazing jump for such a tiny creature. Some fleas can jump over a hundred times their own body length. If people could do that, we would all be superheroes! A 6-foot-tall person would be able to jump 600 feet, the length of two football fields. Those strong flea legs can keep right on jumping, too. Fleas can jump several times a minute for hours on end.

⇐ This picture shows a *cat flea* jumping in slow motion.

What Are Baby Fleas Like?

Most female fleas lay their eggs on their host or in the animal's bedding or nest. Eggs laid on the host often fall off where the animal sleeps. When the eggs hatch, they turn into eyeless, wormlike **larvae.** Flea larvae live in places like animal nests, dirty rugs, and the bedding of pets and farm animals. There they find plenty of food scraps, hair, and droppings to eat. The larvae eat and grow for a few days, weeks, or months. Then they spin a silky case called a *cocoon* around themselves and turn into adults.

This flea cocoon is deep in a family's carpet. ⇒

New adults want to eat as soon as they come out of their cocoons. Even inside the cocoon, a flea can sense whether there is a host nearby. It waits to smell an animal, feel its body heat, or feel it moving nearby. If the flea cannot sense a host, it stays inside its cocoon—sometimes for months.

As soon as a host arrives, the flea leaves its cocoon and heads for its dinner. Long ago, people thought fleas arose magically from lifeless dirt and dust. What they really saw were hungry fleas emerging from their tiny cocoons. Once they become adults, some fleas can live for a year or more.

⇐ This *rat flea* is feeding on a person.

What Do Fleas Eat?

Flea larvae eat a variety of things, from hair to animal droppings. But adult fleas eat only one thing—blood. They are very good at finding host animals. They are most likely to feed on animals that stay in the same place night after night. That is because eggs laid in the animal's sleeping place can find their host very easily. Most animals that wander from place to place are not as badly bothered by fleas.

Here a cat flea's mouth parts are piercing the skin of its host. ⇒

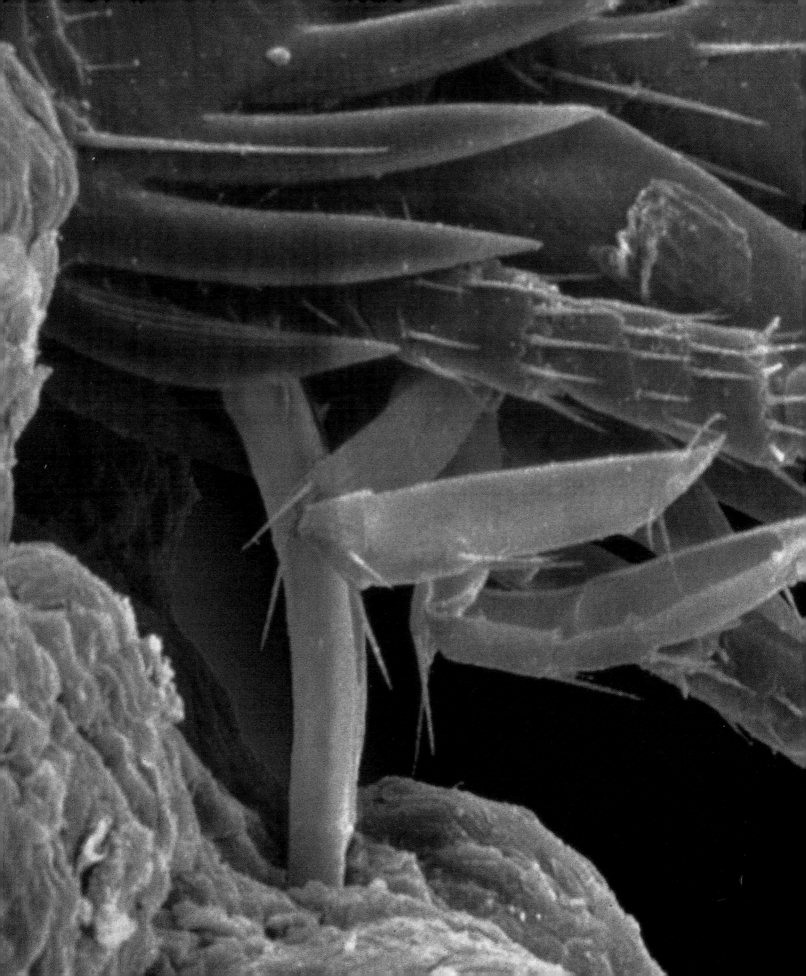

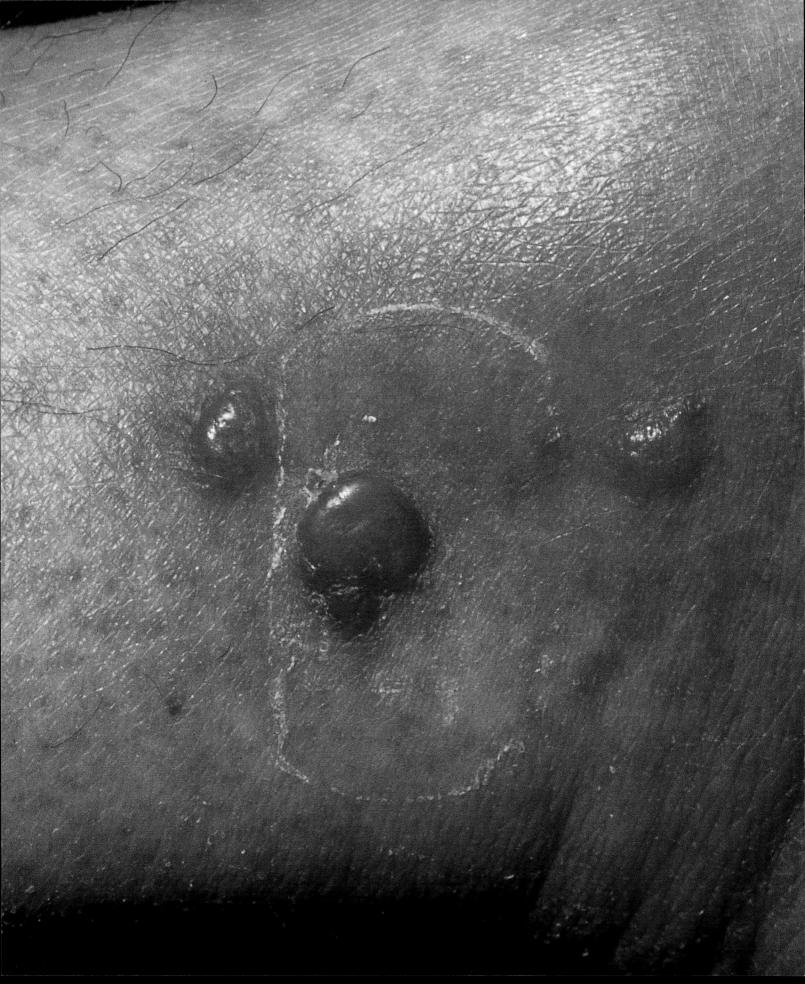

Are Fleas Dangerous?

It's hard to imagine that fleas could be dangerous. After all, they are so tiny! But flea bites itch and can get swollen and infected. Fleas also carry germs and spread disease. Sometimes they carry a deadly disease called *typhus* (TY–fuss). They also spread other parasites. Dog fleas, for example, can spread *tapeworms* that live and grow in the dog's intestines.

⇐ The bumps on this person's skin are itchy flea bites.

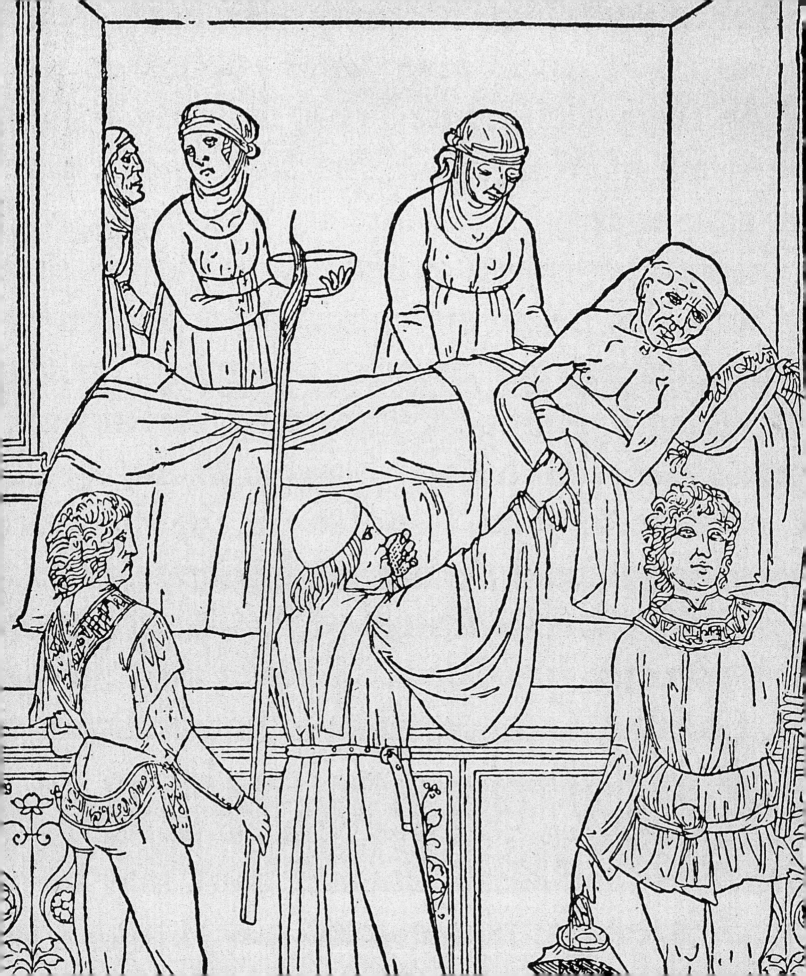

In the Middle Ages, *Oriental rat fleas* spread a terrible disease called **bubonic plague,** or Black Death. Rats were the main carriers of the disease. Fleas bit the infected rats and then bit people, passing on the disease.

A fast-spreading disease that affects large numbers of people is called an **epidemic.** From the 1300s through the 1600s, Black Death epidemics killed half the people in Europe. Even today, people sometimes get sick from bubonic plague.

⇐ This drawing shows a wealthy man (on the bed) who is dying of bubonic plague. His doctor is holding his hand.

How Can Fleas Be Controlled?

Many people see fleas only when their dogs or cats bring them in from outside. One of the simplest ways to prevent fleas is to keep houses, rugs, pets, and bedding nice and clean. Then the larvae have little food, and the fleas, eggs, and larvae have few hiding places. If fleas get to be a problem anyway, animal doctors called **veterinarians** can help. New medicines can kill fleas without harming dogs and cats.

This veterinarian is checking a puppy for fleas. ⇒

Fleas have been around for a long, long time. We might not like them, but they have adapted successfully to life all over the world. These pests will be with us for a long time to come, so we must find a way to live with them. Making our homes less appealing to fleas is a good place to start!

Glossary

bubonic plague (boo–BON–ik PLAYG)
The bubonic plague, or Black Death, was a disease that killed millions of people during the Middle Ages. Fleas carried the disease from rats to people.

epidemic (eh–pih–DEH–mik)
An epidemic is a sickness that spreads rapidly through a large group of people. Fleas have caused many serious epidemics throughout history.

hosts (HOHSTS)
Animals that fleas and other parasites live on are called hosts. Fleas live on a wide variety of hosts.

larvae (LAR–vee)
Larvae are baby insects that look like little worms. Baby fleas live as larvae for a few days, weeks, or months before turning into adults.

parasites (PAYR–uh–sites)
Parasites are animals that feed on other living animals. Fleas are parasites.

species (SPEE–sheez)
A species is a different kind of an animal. There are about 16,000 species of fleas in the world.

veterinarians (veh–ter–ih–NAYR–ee–unz)
Veterinarians are doctors who treat dogs, cats, and other animals. Veterinarians can provide medicines that kill fleas safely.

Web Sites

Learn more about fleas:

http://kidshealth.org/kid/games/flea.html

http://www.orkin.com/fleas/fleasindex.html

www.ag.ohio-state.edu/~ohioline/hyg-fact/2000/2081.html